Josefa Castillo
Erica Coronado
Carlos Jacomino

FOUNDATIONS OF RESILIENCE

Josefa Castillo
Erica Coronado
Carlos Jacomino

FOUNDATIONS OF RESILIENCE

HIDDEN HISTORIES OF STEEL AND STONE IN CIVIL ENGINEERING

ScienciaScripts

Imprint

Cover image: www.ingimage.com

This book is a translation from the original published under ISBN 978-613-9-40001-0.

Publisher:
Sciencia Scripts
is a trademark of
Dodo Books Indian Ocean Ltd. and OmniScriptum S.R.L publishing group

120 High Road, East Finchley, London, N2 9ED, United Kingdom
Str. Armeneasca 28/1, office 1, Chisinau MD-2012, Republic of Moldova, Europe
Managing Directors: Ieva Konstantinova, Victoria Ursu
info@omniscriptum.com

Printed at: see last page
ISBN: 978-620-8-38164-6

For the reader:

This book takes you on a fascinating adventure through the most iconic works of civil engineering. Here you will learn the hidden stories of structures that have defied time, from the ancient pyramids of Egypt to Roman aqueducts. Each chapter explores a key aspect: from the earliest materials and techniques, through the most enduring monuments, to the principles that still inspire modern engineering.

Get ready to discover how these wonders were built with a blend of science, creativity and an unstoppable will to transcend. "Foundations of Resilience" not only tells you the story of these structures, but also of the human ingenuity and effort that made them possible. Dare to explore the legacy that connects us to the past and inspires us towards the future!

Index

Introduction

More than 400,000 years ago, in a corner lost in time, the first humans began to build shelters, marking the beginning of an extraordinary architectural journey. This journey intensified 12,000 years ago with the shift to sedentism, when humankind began to erect permanent dwellings and shape the first settlements. The first cities with streets, drainage systems and public spaces emerged, reflecting the complexity of a growing society and its desire to make its mark.

Each structural collapse was a lesson that prompted new techniques, from the transition from wood to stone to the use of precise tools. These advances were the basis for constructions that not only defied time, but also symbolised the collective identity of their creators. Thus, in every stone and every structure, we discover how the human yearning to transcend has shaped our world and connects us deeply to the roots of our shared history.

CAPÍTULO I:
Nacimiento
DE LA
Ingeniería
Civil

Evolution of early human constructions

More than 400,000 years ago, our ancestors took the first steps in what would become one of humanity's most significant feats: the construction of shelters (Ballantyne, 2019). Beginning with simple structures of branches and skins, these early builders laid the foundations of what would eventually become modern civil engineering. This process was not only an act of survival, but a testament to human ingenuity and the ability to adapt to a changing environment. Natural caves served as the first laboratories of human construction. According to Moropoulou (2020), in these spaces, our ancestors learned fundamental principles of protection from the elements, ventilation and space management. The modifications they made to these natural caverns - enlargements, divisions of space and the creation of specific areas for different activities - represented the first exercises in architectural design. This ingenious use of space not only provided shelter, but also allowed for the development of social and cultural activities within these protected environments.

The real quantum leap in construction came with the transition from nomadism to sedentism, approximately 12,000 years ago (Smith, 2019). This fundamental transformation in the human way of life catalysed the need for permanent structures. Agriculture emerged as a vital source of sustenance, leading communities to settle in a fixed location. The first entirely human-built dwellings appeared in the form of half-buried circular huts, using local materials such as wood, thatch and mud. These huts not only provided shelter, but were also adaptable to changing seasons and climatic conditions. The Çatalhöyük settlements in present-day Turkey (7500 BC) represent one of the oldest and best-preserved examples of this evolution (Hodder, 2020). According to the Çatalhöyük Research Project (2022), these constructions already showed a sophisticated understanding of spatial organisation and the use of composite materials, such as adobe and the tapial technique. The dense arrangement of dwellings and communal spaces reveals a conscious planning that encouraged social interactions and trade among the inhabitants.

With the growth of settlements, the need for planning arose. Lendering (2021) identifies that early Mesopotamian cities introduced revolutionary concepts such as planned streets, drainage systems, defined public spaces and building hierarchy. Uruk, considered the first mega-city in history (4500 BC), demonstrated how building evolution went hand in hand with social organisation (Fagan, 2021). Its imposing walls and temples not only served functional purposes, but also represented the earliest examples of monumental construction. These structures reflected not only political and religious power but also a collective sense of identity among their inhabitants.

The process of constructive evolution was not without its failures. Menon and Radhakrishnan (2022) point out that every structural collapse, every defect in materials and every drainage problem became a vital lesson. Ancient builders developed an empirical method based on trial and error that, although slow, proved tremendously effective. This approach allowed civilisations to learn about the physical and mechanical properties of available materials. A notable example is the evolution of construction techniques in ancient Egypt. Kostof (2023) describes how early pyramids, such as the Step Pyramid of Djoser (2650 BC), show clear evidence of adjustments and corrections to their design, the product of learning accumulated from previous constructions. Constant adaptation over time allowed not only for improvements in structural quality but also for innovations in architectural techniques.

Nuttgens (2021) emphasises how many of the basic principles discovered then are still relevant today. These include the importance of a solid foundation to ensure structural stability;

the need for adequate ventilation to maintain healthy internal conditions; the proper management of loads and distribution of forces to prevent collapse; and the appropriate selection of materials according to climate to ensure durability and comfort. This knowledge, acquired through millennia of experimentation and innovation, is the foundation on which all modern civil engineering building was built.

The materials revolution: from wood to stone

The historical transition from the use of wood to stone represents one of the most significant developments in the history of construction, marking a radical change in building techniques and in the durability and scope of human structures. This process not only transformed the way living spaces were built, but also had profound social, economic and cultural repercussions. During the early stages of human construction, wood dominated the building landscape for fundamental reasons. Its easy availability in most regions, its simplicity in processing and handling, its relatively light weight for transport and its versatility in applications made it a preferred material. Prehistoric communities used wood to build shelters, tools and furniture, taking advantage of its natural properties. However, despite its advantages, wood had significant limitations. Its vulnerability to fire, deterioration due to exposure to natural elements such as moisture and insects, and its limited ability to withstand great heights or structural spans made its use unsustainable in the long term.

The transition phase towards the use of stone was characterised by a hybrid approach combining wood and stone. During this period, stone foundations began to provide greater structural stability, better protection against soil moisture and an increase in the overall durability of buildings. Mixed systems included stone walls with timber roofs, stone columns with timber beams and the development of anchoring techniques between these disparate materials. This combination allowed builders to take advantage of the best of both worlds: the strength and durability of stone together with the lightness and ease of handling of timber.

The final transition to stone construction marked a turning point in ancient architecture. Technical advances during this period included the development of sophisticated stonemasonry techniques, innovations in lifting and transportation systems, as well as the creation of new specialised tools. Developments in bonding and mortar techniques were also crucial; builders began to experiment with different types of mortar that improved the cohesion between stones. The structural advantages of the exclusive use of stone were evident: higher load-bearing capacity, better fire resistance and superior durability enabled taller and more complex structures to be built.

This change had profound social implications. With the increase in building complexity came the emergence of specialised guilds of stonemasons who mastered the art of stonework. These guilds were not only responsible for the technical aspect, but also developed more complex organisational systems of work. The need for precise measurement and spatial planning led to significant developments in mathematics and applied geometry, which enabled architects to design more ambitious buildings. In addition, changes in construction methods had a direct impact on urban planning and architectural design. Cities began to be structured around permanent monuments, reflecting not only political power but also a collective sense of cultural identity. Great works of architecture became enduring symbols of human ingenuity.

The resulting technical innovations were numerous and covered different aspects of the building process. In cutting and finishing techniques, specialised tools were developed that allowed for more precise work with the stone. Quarrying methods evolved to maximise efficiency and minimise waste. Polishing and finishing systems also improved significantly, allowing for finer finishes that enhanced the final aesthetics of the product. Bonding systems evolved to include mortarless rigging, as well as the advanced development of mortars and binders that increased cohesion between stones. Mechanical techniques for transport and laying were revolutionised by the use of pulleys and levers, inclined ramps and temporary scaffolding that facilitated working at height.

This revolution in materials laid the foundations for the subsequent development of both architecture and civil engineering, leaving a legacy that continues to influence modern building practices. The transition from the predominant use of timber to stone not only transformed the way we build; it also altered our conception of permanence and monumentality in architecture. The tangible impact of this change can be seen in the ancient structures that have survived to this day. From Greek temples to medieval cathedrals, these buildings are not only testaments to human ingenuity but also clear examples of the fundamental impact this material revolution had on civilisational development.

Development of fundamental tools and techniques

The earliest manifestations of civil engineering arose from the human need to modify their environment in order to survive and prosper. This evolution began with rudimentary tools that gradually evolved into more sophisticated and precise instruments. At the dawn of civilisation, graduated strings became the first precise measuring tools. The Egyptians, faced with the annual challenge of re-establishing land boundaries after the Nile floods, developed a system of measurement based on the royal cubit, a standardised unit that enabled them to achieve astonishing levels of accuracy in their constructions. The Great Pyramid of Giza, with a deviation of less than 1 degree in its base angles, testifies to the effectiveness of these methods.

Plumb bobs, apparently simple but incredibly effective, revolutionised vertical construction. This instrument, consisting of a weight attached to a rope, harnessed gravity to provide a perfectly vertical line, proving essential in the construction of tall structures and in checking the verticality of walls and columns. The transition from the Bronze Age to the Iron Age marked a turning point in the development of construction tools. The durability and strength limitations of bronze tools were overcome by the new iron tools, allowing harder materials to be worked with greater precision. Chisels were diversified according to their function: pointed for initial work, flat for finishing and serrated for specific textures.

In terms of construction techniques, the evolution was gradual but steady. Ancient builders developed sophisticated methods for foundations, beginning with careful observation of the terrain. Mesopotamian civilisations, for example, devised advanced techniques for building in marshy terrain, using interwoven reed mats and alternating layers of bricks to create stable foundations, a technique fundamental to the construction of their imposing ziggurats. The Romans marked a turning point with the development of opus caementicium, their version of concrete. This innovation allowed the construction of larger and more complex structures, the Pantheon in Rome, with its impressive 43.3 metre dome, being the most notable example of its constructive potential.

Figure 1
Image: This is the opus caementicium

Source: Shutterstock

Figure 2
Image of the Pantheon in Rome

Source: National Geographic History

Transporting heavy materials represented one of the greatest challenges. The Egyptians developed ingenious systems for moving multi-ton blocks of stone using wooden sledges on wet sand to reduce friction. The Greeks and Romans, meanwhile, perfected pulley systems, developing cranes such as the 'Trispastos', which reduced the effort needed to lift weights by a third. The organisation of work also underwent a significant evolution. Clear hierarchies of labour were established, with master craftsmen supervising apprentices and workers. This structure allowed for specialisation and the effective transmission of knowledge, contributing to the continuous improvement of building techniques.

Drainage and water management systems were another crucial area of innovation. Ancient builders developed sophisticated canal, aqueduct and sewerage systems. The Romans, in particular, excelled in this field by creating networks of aqueducts that still amaze today for their engineering and durability. The documentation and transmission of construction knowledge also played a fundamental role. Although initially knowledge was mainly transmitted orally and practically, the first treatises and manuals on construction gradually emerged. Vitruvius, with his "Ten Books of Architecture", provides one of the most complete testimonies on building techniques in antiquity.

These technical and methodological innovations laid the foundations for modern civil engineering. Accuracy in measurement, efficiency in the transport of materials, organisation of work and documentation of knowledge were crucial elements that enabled the development of increasingly ambitious and complex projects. The importance of these fundamental tools and techniques lies not only in their immediate practical utility but also in how they transformed the way humans conceived and executed their construction projects.

Each innovation opened up new possibilities and posed new challenges, driving a continuous cycle of improvement and development that continues to this day. For example, the use of iron not only improved the tools used in construction, but also facilitated advances in other areas such as agriculture and metallurgy. Metal tools enabled ancient societies to expand their productive capacities and improve their quality of life.

Early measurement and design systems

The history of civil engineering is intrinsically linked to the development of measurement and design systems, fundamental elements that have enabled civilisations to build complex and functional infrastructures. Since prehistoric times, humans have felt the need to measure in order to satisfy their basic needs, such as shelter construction, hunting and agriculture. Early units of measurement were informal and based on parts of the human body, such as the foot or the elbow. However, these measurements were highly subjective and varied between individuals, leading to confusion in trade and construction activities.

As civilisations advanced, more organised systems of measurement were established. In Egypt, for example, the 'royal cubit' was used as the standard unit for building monuments, including the pyramids. This system represented a significant advance over previous informal measurements, although it still had limitations in terms of accuracy and standardisation. The Babylonians introduced a sexagesimal system (base 60), which still influences how we measure time and angles today; for example, 60 seconds in a minute and 360 degrees in a circle. The Greeks also made important contributions to the development of geometry, laying the foundations for modern architectural design. The Romans, meanwhile, developed a more structured system that included units such as the Roman mile and the Roman foot. Their practical approach to construction led to the creation of durable infrastructure such as aqueducts and roads. Roman engineering was notable not only for its scale but also for its attention to detail in planning and execution, which required precise measures to ensure the stability and functionality of its works.

A crucial milestone in the history of measurement was the French Revolution, which led to the creation of the decimal metric system in 1795. This system was designed to be logical and universal, based on units that could be accurately replicated. The metre was initially defined as a fraction of the earth's meridian, providing a scientific basis for its use. The decimal structure of the metric system greatly facilitated the work of civil engineers by allowing more accurate and standardised calculations. The metric system is characterised by its decimal organisation, meaning that units are arranged in powers of ten. This simplified conversions between different quantities; for example: 1 kilometre (km) = 1,000 metres (m), 1 hectare (ha) = 10,000 square metres (m^2), and 1 litre (L) = 1,000 millilitres (mL). This decimal logic not only facilitated the daily work of engineers, but also improved communication between professionals in different regions.

The International System of Units (SI), formalised in 1960 and most recently revised in 2019, represents a significant advance in the global standardisation of measurements. In this revision, all SI units were redefined from fundamental constants of nature, eliminating any dependence on specific material artefacts. The "SI" is based on seven fundamental units: metre (m) for length, kilogram (kg) for mass, second (s) for time, ampere (A) for electric current, kelvin (K) for temperature, mole (mol) for amount of substance and candela (cd) for luminous intensity. These units allow a wide range of applications in civil engineering, from structural design to thermal and environmental analysis.

The implementation of the metric and later the "SI" system has transformed how civil engineering projects are carried out. Technical specifications now require that all documents, from drawings to technical reports, use metric units exclusively. This not only improves clarity, but also significantly reduces errors arising from incorrect conversions between systems. Standardisation has also led to the creation of international standards governing various aspects of design and construction. Organisations such as ISO (International Organisation for Standardisation) have developed standards that ensure quality and safety in projects globally. These standards range from materials to construction procedures, ensuring that all engineers work under common criteria.

In addition, the unique use of SI allows for better technology integration. With the rise of advanced software used in computer-aided design (CAD) and building information modelling (BIM), engineers can work with a consistent set of units that are internationally recognised. This optimises not only the structural design process but also logistical planning and project lifecycle management.

Despite the obvious benefits of IS, many countries still face challenges in its full implementation. In some regions, practices based on traditional or local systems persist and may hinder full adoption. Cultural resistance to change is one of the main obstacles; many practitioners are used to working with imperial or local systems and may be reluctant to change due to the learning curve associated with new units and methods. In the United States, for example, although the use of metric has been promoted for decades, many industries continue to use imperial measurements due to their cultural and historical roots. This creates complications when working with international partners or when importing/exporting products.

The future of civil engineering will increasingly depend on effective integration of IS with emerging technologies such as artificial intelligence (AI), predictive analytics and automated construction using techniques such as 3D printing. These innovations will not only increase efficiency, but also enable a more sustainable approach to urban development. For example, the use of smart sensors can provide real-time data on the structural condition of a building or bridge, enabling proactive interventions before significant failures occur.

Universal building principles that endure

Civil engineering has been a fundamental pillar in the development of human societies, and throughout its rich history it has been guided by a set of universal principles that have endured and evolved over time. These principles not only reflect the accumulated experience of past civilisations, but also address human needs and the challenges of the built environment. The following is an examination of these fundamental principles that continue to influence modern building practices.

The principle of utility is essential in construction, as it states that every structure must be fit for the purpose for which it was designed. From the earliest dwellings to contemporary architectural complexes, functionality has been a key criterion in design. This involves not only the efficient arrangement of spaces, but also accessibility and ease of use. Functional design considers how occupants interact with the space, ensuring that each area fulfils its specific purpose. For example, in public buildings such as schools and hospitals, the organisation of space can directly influence service efficiency and user experience. Functionality also extends to the adaptability of spaces. As the needs of communities change, structures must be able to adjust to new uses without requiring a complete refurbishment. This flexible approach not only maximises the value of built space, but also contributes to more efficient resource management.

Sustainability has emerged as a fundamental principle in modern construction, especially in response to today's environmental challenges. This concept involves designing and building in a way that minimises environmental impact and maximises resource efficiency. Sustainable construction seeks not only to reduce energy consumption during building operation, but also to consider the full life cycle of the materials used. The implementation of technologies such as solar panels, rainwater harvesting systems and recyclable materials are clear examples of how this principle is applied in contemporary practice. In addition, energy efficiency not only reduces long-term operating costs, but also contributes to a healthier environment for occupants and the planet.

Durability is another key principle that refers to the ability of a structure to withstand the forces of time and adverse environmental conditions. From ancient Egyptian pyramids to modern skyscrapers, strength has been a determining factor in architectural design. Structures must be able to withstand static and dynamic loads, as well as resist natural disasters such as earthquakes and floods. This principle ensures not only the safety of the occupants, but also minimises the costs associated with repairs and maintenance over time. The right choice of materials and construction techniques is crucial to ensure the longevity of any building. For example, the use of reinforced concrete or structural steel can significantly increase the strength and durability of a structure.

Adaptability is an essential principle in contemporary construction. As societies evolve, so do their functional needs. Designing buildings with flexibility allows them to be used for different purposes over time without the need for a complete refurbishment. For example, a space initially designed as an office can easily be transformed into a residential or commercial area if properly planned from the outset. This ability to adapt to future changes not only maximises the value of the built space, but also contributes to more efficient resource management. The implementation of modular spaces or mobile systems within buildings can facilitate this flexibility.

Aesthetics is a fundamental aspect of construction that must be taken seriously. Structures should integrate harmoniously with their cultural and natural surroundings. This principle involves not only architectural form and style, but also how these elements respond to the historical and social context of the place where they are located. Architecture should reflect the cultural identity of a community and contribute to the urban landscape without clashing with it. Landmark buildings such as museums or libraries can become cultural symbols that enrich their surroundings and inspire future generations.

Innovation has been a key driver in the development of new construction methods and materials throughout history. From the invention of concrete to the current use of advanced technologies such as BIM (Building Information Modelling) and 3D printing, each technical advance has led to improved efficiency, precision and sustainability in construction projects. The adoption of new technologies not only optimises construction processes, but also opens up new creative possibilities for architects and engineers. For example, the use of advanced software enables accurate simulations that help to foresee problems before actual construction begins.

Finally, another fundamental principle is the importance of proper documentation and regulatory compliance in all aspects of the construction process. Standards set minimum standards to ensure the structural safety, functionality and sustainability of buildings. Clear documentation allows for better communication between all actors involved in a project - from architects to contractors - ensuring that everyone is working towards the same common goal.

Universal building principles that endure are more than just guidelines; they are essential fundamentals that have guided civilisations throughout time and continue to shape our built environment today. By integrating these principles into every phase of the building process - from initial design to final execution - you can ensure not only the quality of the end result, but also its long-term relevance in a changing world.

These principles are testament to human ingenuity and reflect our ability to adapt to new realities while honouring the lessons learned from the past. As we move into the future, it is essential to continue to apply these fundamentals to build more sustainable, inclusive and resilient environments that meet both the current and future needs of our societies.

CAPITULO
II
MONUMENTOS
DE ETERNIDAD

The Great Pyramid of Giza: Precision and Geometry

The Great Pyramid of Giza, built between 2580 and 2560 BC, stands as the largest of the three pyramid monuments located at Giza and is the only one of the Seven Wonders of the Ancient World still standing. This impressive architectural landmark was erected for the pharaoh Cheops, also known as Khufu, and is a testament to the extraordinary skill of the ancient Egyptians in mathematics, astronomy and architecture. Originally, the pyramid reached a height of 146.6 metres, making it the tallest building in the world for more than 3,800 years, a record that was not broken until the construction of Ulm Cathedral in Germany in 1890. Its base occupies more than 13 acres and is composed of approximately 2.3 million limestone blocks, each weighing between 2.5 and 15 tons.

Figure 3
Image of the amazing pyramids of Giza

Source: Pinterest

Astronomical Alignment and Geometric Design

One of the most fascinating aspects of the Great Pyramid is its near-perfect alignment with the cardinal directions. With a deviation of only 0.067 degrees, this precision suggests a profound knowledge of astronomy on the part of the Egyptian architects. The ancient Egyptians used a system of astronomical observation to ensure that the pyramid was oriented correctly. It is believed that they used tools such as the gnomon, a stick that casts shadows, and other rudimentary instruments to measure the position of the sun and stars. This approach not only demonstrates their technical skill, but also their belief in the importance of worldview in architecture. The pyramid was not only a mausoleum, but also a symbol of the connection between heaven and earth.

The design of the pyramid is not only a feat of construction; it also has a deep symbolic meaning. The pyramid shape represents the rays of the sun, reinforcing the pharaoh's connection to Ra, the sun god and one of the most important deities in the Egyptian pantheon. The pyramid functioned as a bridge between earth and heaven, facilitating the pharaoh's ascent to the afterlife. This spiritual link is manifested in the religious and funerary texts that adorn the walls of the inner chambers, where the deities are invoked to protect the king on his journey to the afterlife.

Funerary Complex and Symbolism

The funerary complex surrounding the pyramid includes temples, other pyramids and tombs, all designed to ease the pharaoh's transition to the afterlife. The importance of the afterlife in Egyptian culture is reflected in the elaborate funerary rituals and the construction of richly decorated tombs. The use of hieroglyphs and reliefs on the pyramid walls and in the burial chambers was essential to ensure that the pharaoh had everything he needed in his afterlife. Inside the pyramid, chambers have been found that contained treasures, statues and other objects that it was believed the pharaoh would need in the afterlife, indicating the wealth and power he possessed during his lifetime.

The funerary complex also includes the famous Sphinx of Giza, which is believed to have been built during the same period and represents the strength and wisdom of the pharaoh. The Sphinx, with its lion-like body and human head, symbolises the protection and vigilance of the pharaoh, adding another level of significance to the complex.

Construction methods

The construction of the Great Pyramid has been the subject of study for centuries, and the methods used to move and place the massive blocks of stone have generated numerous theories. Egyptologists have proposed that workers used ramps and counterweights, as well as a system of levers to facilitate the transport of the stones from the quarries to the construction site. Some theories suggest that straight ramps were used, while others indicate that the ramps may have been spiral, allowing easier access to the higher sections of the pyramid.

This innovative approach to engineering is a testament to the organisation and skill of the workforce involved in the project, which is estimated to have consisted of thousands of workers, including craftsmen, engineers and labourers. Although some popular narratives suggest that slaves built the pyramid, archaeological evidence indicates that these labourers were mostly peasants who worked during the floods of the Nile, when they were unable to cultivate their land. Records indicate that these workers were well fed and received medical care, suggesting that their work on the pyramid was considered a sacred and honourable service.

Resilience over the years

Over the centuries, the Great Pyramid has withstood numerous challenges, from natural disasters to human intervention. Its robust construction, based on advanced techniques and the use of local materials, has allowed this monumental structure to survive earthquakes and erosion. Despite the earthquakes that have affected the region, the pyramid has maintained its structural integrity thanks to its earthquake-resistant design. The pyramid shape is not only aesthetically pleasing, but also provides stability, distributing weight evenly and minimising the risk of collapse.

During the Middle Ages, the pyramid was looted and plundered. Stones from the pyramid were reused in the construction of mosques and other buildings in Cairo. However, despite this damage, the main structure of the pyramid has remained standing. Restoration efforts began in the 19th century, and since then it has been the subject of major archaeological and conservation studies.

In the 20th century, the Great Pyramid became a symbol of ancient Egypt and a global tourist destination. In 1979, it was declared a UNESCO World Heritage Site, which highlighted its cultural importance and has promoted conservation efforts. Over the years, various preservation techniques have been implemented to protect the structure from erosion and the impact of tourism, including the control of environmental conditions at the site.

Legacy and Significance

The Great Pyramid is not only a testament to the power and wealth of Egyptian civilisation, but also an enduring symbol of the human capacity to create and build. Over the centuries, it has been an object of awe and mystery, inspiring explorers, archaeologists and scholars. Its influence extends beyond ancient history; it has become a symbol of engineering and architecture, proving that human ambition can transcend time and space.

The pyramid remains a major tourist destination, attracting millions of visitors each year who seek to experience this wonder of the ancient world up close. In addition, its image and history have been depicted in numerous works of art, literature and film, making it a global cultural icon. The Great Pyramid of Giza is not just a monument, but a legacy that continues to inspire awe and wonder around the world.

Its ability to stand the test of time and remain a beacon of ancient civilisation is a testament to human ingenuity and the eternal desire to seek immortality through architectural creation. In an ever-changing world, the Great Pyramid of Giza remains a symbol of stability, continuity and mankind's relentless quest to leave a lasting mark on history.

The Roman Colosseum: The Revolution of the Arch and the Vault

The Roman Colosseum, inaugurated in 80 AD under Emperor Titus, is one of the greatest architectural achievements of Roman civilisation and an enduring icon of Western culture. This impressive amphitheatre, originally known as the Flavian Amphitheatre, reflects not only Rome's greatness, but also its ability to create massive and functional spaces that have withstood the test of time. Able to hold between 50,000 and 80,000 spectators, the Colosseum was the setting for numerous public events, including gladiatorial combats, theatrical performances and other spectacles that kept the populace entertained.

Architectural Innovations: Arches and Vaulting

One of the most notable innovations of the Colosseum is its extensive use of arches and vaults, which revolutionised the architecture of the time. The construction of the Colosseum was based on a system of round arches, which distributed the weight of the structure more efficiently, allowing the creation of large interior spaces. This technique not only provided stability, but also allowed for an attractive and functional design.

Structure and Design

The structure of the Colosseum is composed of a system of levels that use arches to create aisles and seating. Each level is designed with arches that form a support system, allowing the tiers to rise on several levels, ensuring that all spectators have a clear view of the spectacle. The use of vaults in the construction also made it possible to create underground spaces, known as hypogeum, which served as storage and preparation areas for events.

The Colosseum is an example of architectural ingenuity that combines functionality and aesthetics. Its design includes a longitudinal axis that connects the main entrances to the central area, facilitating the movement of large crowds. This focus on circulation and accessibility is a testament to the careful planning that characterises Roman engineering.

The Colosseum also had a roofing system that could be deployed to protect the audience from inclement weather, a further testament to the sophistication of Roman engineering. This roof, known as a *velarium*, was operated by sailors who used a system of ropes and pulleys to spread the fabric over the areas occupied by the audience, thus providing shade and protection.

In addition, Roman engineers implemented innovations such as emergency exits and rapid access systems, demonstrating an advanced approach to public safety and comfort. With its 80 entrances, the Colosseum's design allowed for efficient access, minimising the risk of crowding and ensuring rapid evacuation in the event of an emergency. This level of planning reflects a genuine concern for the spectator experience as well as for spectator safety at mass events.

Culture and Entertainment in Rome

The Colosseum was not only a space for entertainment, but also an epicentre of Roman public life, where events were held that united the community around grandiose spectacles. Gladiatorial combats, which often resulted in the death of the participants, were a form of entertainment that reflected the brutality and power of Roman society. These spectacles not only served to entertain, but were also a form of political propaganda, showing the greatness of the empire and the benevolence of the emperor.

The events in the Colosseum went beyond gladiatorial combat. There were animal hunts, executions, re-enactments of naval battles and even theatrical performances. This diversity of spectacles reflected the complexity of Roman culture and its desire to impress both citizens and visitors.

The connection between political power and entertainment was palpable. Emperors used the Colosseum as a tool to gain popularity and support among the population, organising games and shows that were often free of charge. This political strategy ensured mass participation of the population, giving citizens a sense of belonging and pride.

Resilience and Conservation through the Centuries

Despite damage over the centuries, the Colosseum remains a powerful reminder of human ingenuity and the ability to create structures that defy time. From earthquakes to looting, the Colosseum's history has been marked by challenges, but its resilience is remarkable.

Natural Disasters

The structure has survived several devastating earthquakes, including those of 847 and 1231 AD, which caused significant damage but did not completely destroy the building. The architectural form of the Colosseum, with its structure of arches and vaults, gives it an inherent stability that has enabled the building to withstand seismic stresses. The vaults, in particular, distribute the weight evenly, which helps prevent collapse during seismic events.

Looting and Reuse of Materials

During the Middle Ages, the Colosseum was used as a dwelling and workshop. Many of its materials were looted for other buildings in Rome, but the fundamental sections of the structure remained standing. In the 18th century, the Colosseum was consecrated as a sacred place, which contributed to its preservation. The Catholic Church, recognising its historical and cultural importance, forbade the demolition of the monument.

Modern Restorations

In the 19th century, restoration efforts began, which included stabilising damaged structures and cleaning the stonework. These works have allowed the Colosseum to continue to be an attraction for millions of visitors, who seek to experience the living history of Rome. The interventions have been carefully planned to maintain the integrity of the monument, using techniques that respect Roman building traditions and the original materials.

Today, the Colosseum is the subject of a rigorous conservation programme that includes monitoring its structural condition and controlling erosion caused by pollution and the wear and tear of time. These actions are essential to ensure that this architectural marvel remains a symbol of civilisation.

Legacy and significance

The Colosseum is not only a symbol of Rome's greatness, but also a testament to man's ability to create spaces that reflect his culture and values. Over the centuries, it has been the subject of admiration and study, inspiring the construction of numerous stadiums and amphitheatres around the world. Its influence extends far beyond antiquity and has established itself in popular culture, appearing in films, literature and as a symbol of civilisation.

Today, the Colosseum is a World Heritage Site and among the New Seven Wonders of the World, attracting millions of visitors each year. Its ability to stand the test of time and remain a beacon of ancient civilisation is a testament to human ingenuity.

The Roman Colosseum, with its impressive history and innovative design, remains an icon of Western culture and a place where time seems to stand still. Each visit allows tourists and scholars to connect with a glorious past, where every stone tells a story of greatness, struggle and endurance.

The Great Wall of China: adaptation to terrain and resilience

The Great Wall of China is a vast system of fortifications stretching more than 21,000 kilometres across northern China, and is considered one of the most iconic engineering and architectural achievements of Chinese civilisation. Its construction began in the 7th century BC and continued until the 16th century, spanning several dynasties, including the Qin, Han and Ming dynasties. This monumental project was designed primarily to protect the borders of the Chinese Empire from external invasion, but also served as a means of communication and control of trade, reflecting the complexity of China's military and cultural history.

The wall was constructed from a variety of materials, including earth, wood, brick and stone, depending on the availability of resources in each region. In mountainous areas, cut stone blocks were used, while in flatter regions, earth was rammed to form solid walls. This adaptive approach not only optimised the use of local materials, but also allowed the wall to integrate effectively into the surrounding landscape.

One of the most remarkable aspects of the Great Wall is its ability to adapt to the terrain. The structure meanders through mountains, deserts and plains, blending seamlessly into the natural landscape. This adaptation maximises its resistance to the elements and provides strategic advantages to defenders. By constructing the wall to follow the curves and contours of the terrain, the architects were able to minimise weak points and maximise height and visibility in elevated areas. In the mountainous regions of Shanxi Province, for example, techniques were used that incorporated the topography, creating walls that aligned with ridges.

The Great Wall not only serves a military function; it is also a symbol of the perseverance and determination of the Chinese people. Its construction required incredible effort, involving millions of workers over the centuries. Many of them faced extreme conditions and dangers, which has led to the wall being steeped in stories of sacrifice and endurance. Popular legend has it that some of the soldiers who died during construction were buried within it, contributing to its mystique and symbolising the human cost of this monumental project.

Today, the Great Wall is a national symbol and tourist attraction, reminding both the Chinese and the world of China's rich history and culture. In 1987, it was declared a UNESCO World Heritage Site, which has ensured its preservation and continued appreciation by tourists and scholars. Despite its resilience, the wall faces significant challenges, including natural erosion, vandalism and the impact of mass tourism. The Chinese authorities have implemented conservation strategies that include the restoration of damaged sections and the promotion of sustainable tourism that respects cultural heritage.

The Great Wall has left an indelible mark on Chinese culture and the global perception of the country. Its presence in popular culture, film, literature and art has contributed to its status as one of the world's most recognisable monuments. In addition to being a symbol of Chinese civilisation, it represents human ingenuity and the ability of civilisations to overcome seemingly insurmountable challenges.

Figure 4
Image of construction techniques on the Great Wall of China

Source: Pinterest

Machu Picchu: high-altitude and earthquake-resistant engineering

Machu Picchu, the famous Inca city in the Peruvian Andes, is an exceptional example of engineering and urban planning. Built in the 15th century under the rule of the Inca emperor Pachacutec, this citadel is located 2,430 metres above sea level and is known for its breathtaking beauty and architectural sophistication. Its isolation and the natural beauty of its surroundings have made it one of the most important tourist destinations in the world, attracting millions of visitors each year.

Figure 5
Image of the view of Machu Pichu

Source: Pinterest

One of the most striking features of Machu Picchu is its earthquake-resistant design. The Incas used advanced construction techniques, such as the use of stone blocks carved to fit together perfectly without mortar. This technique allows the structures to adapt to seismic movements, which has contributed to the preservation of many buildings over the centuries. The walls of Machu Picchu, built with precision-cut stones, have withstood earthquakes that would have toppled buildings from other historical periods.

In addition, the layout of the terraced buildings provided an efficient drainage system, preventing landslides and soil erosion. These agricultural terraces surrounding the city not only provided food, but also helped to stabilise the soil, demonstrating the deep knowledge of agriculture and ecology that the Incas possessed. They grew a variety of crops at different altitudes, ensuring food diversity and community sustainability.

Machu Picchu is considered a sacred site, linked to Inca cosmology. The orientation of many of its structures is aligned with astronomical events, such as the solstices, suggesting that the site had religious and ceremonial significance. The famous Intihuatana, a ritual stone, was probably used for ceremonies related to the sun and agriculture. The strategic location of the city, surrounded by mountains and lush nature, reinforces the spiritual connection the Incas had with their environment.

The city was rediscovered in 1911 by explorer Hiram Bingham, who made it known to the modern world. His account and the images he took to the press generated great interest in the site, making it a symbol of Inca civilisation. Since then, Machu Picchu has been an object of study and admiration, and its recognition as a UNESCO World Heritage Site in 1983 has ensured its preservation for future generations.

The combination of its natural beauty, rich history and ingenious architecture continues to fascinate visitors from all over the world. Machu Picchu is not only a testament to human ingenuity, but also a symbol of Inca culture and its relationship with nature. Its legacy lives on, reminding us of the importance of preserving and appreciating our historical and cultural roots. It has also inspired a renaissance of interest in Inca culture and fostered greater respect for indigenous traditions around the world.

Today, Machu Picchu faces challenges such as mass tourism and climate change, making the need for conservation efforts even more urgent. Machu Picchu's influence transcends its scenic beauty; it represents the ability of civilisations to adapt and thrive in challenging environments, and its history remains a beacon of inspiration for current and future generations.

Loyola College Nicaragua: A Beacon of Education and Resilience

Colegio Loyola de Nicaragua is an outstanding example of educational and religious architecture in the country. Founded in 1914 by the Society of Jesus, this school is based in Managua and has been a fundamental pillar in the academic and moral formation of generations of Nicaraguan students. Its mission focuses on integral education, promoting values such as justice, solidarity and respect, in an environment that fosters personal and community development.

The architectural design of Loyola College reflects the influence of the colonial style, characterised by its use of arches, interior courtyards and an imposing façade that evokes a sense of tradition and solemnity. The structure is surrounded by gardens and open spaces, providing areas conducive to reflection and outdoor learning. This natural setting is integrated into the educational experience, allowing students to connect with their surroundings and cultivate a sense of environmental responsibility.

One of the most outstanding features of the school is its focus on inclusive education and the holistic development of students. It offers a wide range of academic programmes from elementary to high school, with a strong emphasis on values formation and spirituality. Over the years, it has promoted extracurricular activities that encourage leadership, art and sport, contributing to the formation of citizens committed to their community.

The 1972 earthquake was a devastating event that affected Managua and many of its institutions. Despite significant damage to the city, Colegio Loyola was remarkably resilient. The solidity of its construction and the careful planning of its structure helped to minimise the damage. Although some sections of the school required repairs, most of the buildings remained standing, allowing the educational community to continue to function in the midst of the crisis. This resilience is due in part to the construction techniques used in its construction, which prioritised stability and safety.

Loyola College's legacy extends beyond its physical infrastructure. Over the course of more than a century, it has trained leaders in diverse areas, including politics, education, medicine and the arts. Many of its alumni have played significant roles in Nicaraguan society, carrying with them the principles of justice and service that are instilled from an early age.

In the current context, Loyola College faces challenges related to the modernisation of its infrastructure and the need to adapt to a constantly changing educational environment. However, its commitment to academic excellence and value formation remains its main driving force. The educational community works to maintain the school's relevance in a globalised world, without losing sight of its roots and original mission.

Today, Loyola College Nicaragua is more than an educational institution; it is a symbol of the Jesuit tradition in the country and a beacon of hope for future generations. Its history and its impact on Nicaraguan society make it a living monument to education and to the values that have guided its community over the years. The institution remains a testament to the importance of education in building a more just and equitable society, reminding everyone of the need to invest in the future of young people and in human development.

Figure 6
Image of the Loyola School

Source: Loyola Magazine

CAPITULO
III
EL
DOMINIO
DEL
AGUA

Roman aqueducts: the conquest of gravity

Roman aqueducts represent one of the greatest achievements of ancient engineering and an enduring testament to the ingenuity and technical prowess of the Roman Empire (Hodge, 2002). Although the Romans were not the inventors of this technology, already present in civilisations such as the Greeks and Etruscans, they were the ones who perfected it and brought it to an unprecedented scale (Hansen, 2019). This monumental breakthrough began in 312 BC with the construction of the Aqua Appia, which ushered in an era of hydraulic expansion that would radically transform urban life in the empire (Mays, 2010).

The genius of these structures lies in their apparent simplicity: water flowed by gravity from elevated sources to urban centres, maintaining a constant gentle slope of approximately 0.5% to 3% (Wilson, 2008). However, this apparent simplicity concealed a complex engineering work that required precise topographical surveys. Roman engineers used instruments such as the chorobates (water level) and groma to calculate slopes and gradients with astonishing accuracy (Smith, 2018).

The construction of an aqueduct involved a comprehensive system that began with the collection of water by means of diversion dams, weirs or filtering galleries (Malissard, 2001). The water travelled mainly through the *specus*, a covered channel built of stone and lined with *opus signinum*, a special hydraulic mortar that guaranteed waterproofing (Lancaster, 2010). When the terrain required it, the iconic arcades were built: elevated bridges with overlapping semicircular arches that allowed the necessary slope to be maintained while bridging valleys and depressions (Trevor, 2016).

The Romans developed sophisticated construction techniques that combined different types of mortar according to need: *opus caementicium*, *opus quadratum* and *opus reticulatum* (Adam, 2005). Infrastructure included crucial elements such as *piscinae limariae*, *spiramina* and *putei*, as well as innovations such as hydraulic siphons to cross deep valleys and gate systems for maintenance (Frontino and Rodgers, 2004).

As Bruun (2013) points out, the urban distribution system was equally sophisticated. Water reached the *castellum aquae*, from where it was distributed through a network of lead and ceramic pipes to public fountains and important buildings. Water management was strictly regulated, with specific officials (*curator aquarum*) in charge of the control and maintenance of the water supply, as well as a complex system of concessions and water rights.

The impact of aqueducts on Roman society was revolutionary. These systems not only allowed the growth of large cities, but also significantly improved sanitary conditions by providing access to clean drinking water (DeLaine, 2015). Among the most notable examples that have survived to the present day are the Pont du Gard in France, the Segovia Aqueduct in Spain, the Aqua Claudia in Italy and the Valente Aqueduct in Turkey (Taylor, 2020).

The legacy of Roman aqueducts is indisputable. These structures not only facilitated daily life by providing water for domestic consumption, irrigation and public baths, but also symbolised the power of the Roman Empire over nature (Hodge, 2002). Their ability to transport large volumes of water over long distances using gravity alone is an impressive achievement that has endured over the centuries. The Roman aqueducts are an outstanding example of human ingenuity in antiquity. Their construction not only served a utilitarian function, but also served as an expression of Roman dominion over their natural environment. This legacy endures to this day, inspiring modern engineers and architects in their search for innovative solutions in water management and urban infrastructure.

Yerebatan Cisterns: underground storage

The Yerebatan Cistern (Yerebatan Sarnıcı), also known as the Basilica Cistern, is a magnificent work of architecture that was built during the reign of Emperor Justinian I in the 6th century AD. This structure not only represents the culmination of Byzantine ingenuity in water storage and distribution, but also stands as a symbol of the splendour of ancient Constantinople and the technical prowess of its engineers (Mango, 2015). Located in the heart of the city, this impressive underground cistern is an extraordinary testimony to Byzantine hydraulic engineering and the continuation of the Roman legacy in water management, standing out for its innovative design and functionality (Crow et al., 2008).

With monumental dimensions reaching 140 metres in length and 70 metres in width, the Yerebatan Cistern has a capacity of approximately 100,000 cubic metres of water, making it one of the largest underground cisterns in the world (Bardill, 2017). This imposing structure is 9 metres below ground level and is supported by a total of 336 columns, arranged in 12 rows with 28 columns each. Each column is 9 metres high and topped by brick vaults, while the walls surrounding the cistern are an impressive 4 metres thick and have been waterproofed with a special hydraulic mortar to ensure their functionality over time (Çeçen, 2014).

The columns supporting this magnificent structure are particularly remarkable. As Ward-Perkins (2012) notes, they are mostly from ancient temples and exhibit a variety of architectural orders, the most predominant being Corinthian and Ionic. Each column has finely carved bases and capitals that reflect the master craftsmanship of Byzantine stonemasons. The waterproofing system is another crucial aspect of the design; according to studies by Lancaster (2018), this includes walls lined with hydraulic mortar, a brick-paved floor and an elaborate system of elastic joints designed to prevent leaks.

The hydraulic system that feeds the cistern is connected to the famous Valente Aqueduct, which allows for a constant supply of water. In addition, it has sophisticated mechanisms for filtration and sediment control, thus ensuring the quality of the stored water (Crow, 2012). The cistern served multiple vital functions: not only did it store water for prolonged periods of drought, but it also supplied the Great Palace and adjacent areas. In times of conflict or siege, it served as a crucial strategic reservoir to guarantee the water supply to the population (Dark and Harris, 2008).

A particularly notable aspect of the cistern's operation is its operating system, documented by Westbrook (2019). This system included processes to capture both water from the main aqueduct and rainwater. In addition, a rigorous maintenance programme was implemented, including regular cleaning of sediment and constant monitoring of the quality of the stored water. Water was distributed through a complex network of pipes that fed public fountains and important buildings throughout Constantinople.

Among the most prominent artistic elements within the cistern are the famous Medusa column bases. According to Bassett (2015), these bases not only served a structural function, but also served as apotropaic symbols, intended to protect the sacred space. Architectural decoration includes finely carved capitals and reliefs on the columns, as well as numerous marks left by stonemasons and symbols that show the collaboration between different construction workshops.

The current state of the Yerebatan Cistern is excellent, thanks to numerous restorative interventions that have taken place over time. These restorations have been carried out with a careful approach to respect the historical integrity of the monument (Ahunbay, 2016). Modern restorations have included not only the cleaning and consolidation of existing structures, but also the reinforcement of deteriorated elements and adaptations to facilitate public access. This approach has allowed a balance to be maintained between historic preservation and tourist functionality.

In short, the Basilica Cistern is not only an outstanding example of Byzantine architectural ingenuity but also a cultural and historical site that attracts millions of visitors each year because of its unique beauty and fascinating history. Its importance transcends the merely functional; it has become an enduring symbol of human ingenuity and a tangible representation of the cultural and historical splendour that characterised Constantinople during its heyday.

Hydraulic systems of Petra and Machu Picchu

The mastery of water has been a fundamental pillar in the development of civilisations throughout history, and two outstanding examples of this mastery are the hydraulic systems of Petra in Jordan and Machu Picchu in Peru. Both cultures, despite their geographical and temporal differences, developed ingenious techniques to manage water, thus ensuring the sustainability of their communities.

Petra, known as the 'pink city' because of the colour of its rocks, is famous not only for its impressive stone architecture, but also for its advanced hydraulic system. The Nabataeans, who inhabited this region, faced the challenge of an arid and dry environment. To survive, they developed a sophisticated system of water collection and storage that included canals, cisterns and reservoirs. The Nabataeans constructed canals that diverted rainwater into underground cisterns, allowing for the collection of scarce water. These canals were designed to make the most of every drop of rain, using techniques that minimised evaporation and maximised infiltration of water into the ground. In addition, the cisterns were lined with waterproof mortar to prevent seepage and leaks.

A remarkable aspect of Petra's hydraulic system is its ability to manage rainwater in a mountainous terrain. Nabataean engineers designed structures that allowed water to be diverted to cultivable areas, creating an oasis in the middle of the desert. This efficient water management was crucial to Petra's agricultural development and economic prosperity.

Machu Picchu, on the other hand, presents an equally impressive example of hydraulic engineering. The Incas developed intricate systems of aqueducts and canals to collect and distribute water in this iconic citadel (Top Alpaka Travel, n.d.). The aqueducts, constructed of precisely carved stones, captured water from nearby springs and directed it to agricultural terraces and fountains strategically distributed throughout the complex. Hydraulic planning at Machu Picchu was meticulous. The Incas designed carefully placed water sources that channelled the vital liquid through canals to specific areas. This not only guaranteed access to drinking water, but also created relaxing environments such as thermal baths (Top Alpaka Travel, n.d.). In addition, its efficient drainage system prevented the accumulation of water in the structures, protecting them from potential damage (iAgua, n.d.).

The integration of the hydraulic system with the natural topography is another notable aspect. The Incas adapted their architecture to harmonise with the surroundings, minimising environmental impact and respecting natural water cycles (iAgua, n.d.). This approach reflects not only an advanced understanding of water management but also a deep spiritual connection with nature.

Studies by Wright and Valencia (n.d.) highlight that Inca hydraulic systems included complex components such as open ditches that followed contour lines to conduct water into oxbow lakes where it was filtered before emerging as puquios months later. This ingenious design ensured a constant supply during dry periods, allowing for sustainable agriculture.

Both Petra and Machu Picchu are outstanding examples of human ingenuity in the mastery of water. Through their innovative hydraulic systems, these civilisations not only managed to survive in challenging environments, but also flourished culturally. Effective water management was fundamental to their economic and social development, leaving a lasting legacy that continues to inspire current generations in the search for sustainable water management solutions.

Old canals and harbours

Ancient civilisations manifested themselves in a remarkable way through the construction of canals and harbours, which played a crucial role in the economic, social and military development of these societies. These systems not only facilitated the transport of goods, but also ensured access to vital water resources, which in turn encouraged agriculture and trade.

Navigable canals have been used since ancient times to connect rivers, lakes and oceans, enabling the efficient transport of goods and people. An early example is the system of artificial canals created by the ancient Egyptians, which facilitated navigation and trade on the Nile. These canals were essential for agriculture, as they allowed the irrigation of land in arid areas and contributed to the development of a thriving agricultural economy. The ability to control the water of the Nile was fundamental to Egyptian civilisation, allowing for abundant harvests that supported a growing population.

During Roman times, canal construction reached a new level of sophistication. Roman engineers designed extensive canal systems that not only connected different bodies of water, but also facilitated internal navigation in cities such as Rome. These canals were crucial for trade and military transport, allowing for the efficient movement of troops and supplies throughout the empire. The Roman canal network not only improved the logistics of the empire, but also fostered cultural integration by facilitating exchange between different regions (Wikipedia, n.d.).

One of the most prominent examples is the Corinth Canal, which connects the Ionian Sea with the Aegean Sea. This canal was a monumental work that significantly reduced the navigational distance for ships transiting between these two seas. Its opening allowed faster access to Mediterranean trade routes, thus boosting the regional economy and strengthening trade connections between different cultures.

Ancient ports were central to maritime trade and cultural expansion. From the early shelters built by the Phoenicians to the elaborate harbours of Greek and Roman cities, these spaces were vital for commercial exchange and the projection of naval power. A notable example is the port of Wadi al-Jarf in Egypt, dating from around 2500 BC. This artificial harbour facilitated maritime trade and was strategically located near Memphis, making it a key point for ancient Egyptian trade routes (Wikipedia, n.d.). The Phoenicians also played a crucial role in the development of ports, building facilities in cities such as Sidon and Tyre during the 13th century BC, which became important trading centres in the Mediterranean (Prosertek, n.d.).

In Greece, the port of Piraeus stood out as an essential naval base during the classical era. This port not only served as a departure point for military expeditions, but was also a vibrant commercial centre connecting Athens with other regions of the ancient world. The strategic importance of Piraeus was such that it became one of the main ports in the Mediterranean, facilitating both trade and Greek military projection (Wikipedia, n.d.). The construction of the Heptastadion dyke in Alexandria during the 3rd century BC was another significant milestone; this dyke separated two parts of the port and housed the famous Lighthouse of Alexandria, an architectural marvel and one of the oldest lighthouses in the world (Wikipedia, n.d.).

The evolution of the ports continued over time, adapting to the changing needs of maritime trade. The construction techniques used included local rocks and durable materials that ensured resistance to adverse weather conditions. With the advent of new technologies and construction methods during the Middle and Modern Ages, ports were transformed into multimodal complexes integrating various forms of transport and logistics (Prosertek, n.d.). This adaptation not only improved the efficiency of maritime trade, but also allowed port cities to thrive as economic centres.

These infrastructures not only facilitated trade and communication between civilisations, but also played a key role in economic and military development. The legacy of these hydraulic systems survives to this day, reminding us of the importance of water as a vital resource for human survival and prosperity.

Irrigation and flood control techniques

The development of irrigation and flood control techniques has been fundamental to the progress of ancient civilisations, allowing the transformation of arid territories into arable land and the effective management of river floods, laying the foundations for agricultural and urban development (Hassan, 2011). The Mesopotamians developed one of the first complex irrigation systems, using canals extending from the Tigris and Euphrates rivers. The shaduf system, a counterweight structure for raising water, allowed for the irrigation of high ground and became a fundamental technology that subsequently spread throughout the Middle East (Wilkinson, 2013).

In ancient Egypt, the basin system took advantage of the annual flooding of the Nile. Farmers built dykes and canals to retain water and fertile sediment during the flood, creating a natural irrigation system that allowed up to three harvests per year (Butzer, 2019). The Romans refined these techniques through aqueducts to transport water over long distances, distribution systems with adjustable sluice gates, secondary and tertiary canals, and advanced land levelling techniques (Wilson, 2018).

Civilisations developed various methods to control floods, including retaining structures such as dykes, embankments, floodwalls and diversion channels. Drainage systems incorporated drainage channels, retention basins, sluice gates and diversion tunnels (Jansen, 2017). Technical innovations in irrigation systems included water-lifting methods such as waterwheels, Archimedes screws, chain pumps and rocker systems. Distribution was by means of lined canals, raised aqueducts, inverted siphons and terracotta pipes (Shaw, 2020).

The administration of these systems required complex social organisation, including planning through surveying, flow calculations, infrastructure design and maintenance scheduling. Regulation encompassed water rights, irrigation schedules, collective maintenance and conflict resolution (Scarborough, 2016). Different cultures developed techniques adapted to their local conditions: in China, rice terraces, polder systems and diversion canals were prominent, while in pre-Columbian America, Aztec chinampas, Andean camellones and Inca terraces were prominent (Park, 2014).

Irrigation and flood control systems had profound environmental and social effects. Environmentally, they led to landscape modification, alteration of ecosystems, changes in sedimentation patterns and erosion control. Socially, they drove urban development, labour specialisation, the creation of administrative structures and specific legal systems (Miller, 2020).

Water management in ancient societies demonstrates an astonishing technological and organisational sophistication. The techniques developed not only enabled these civilisations to survive and prosper, but also laid the foundations for many modern water management systems. The legacy of these innovations continues to influence today's irrigation and flood control practices, adapting to new challenges and technologies.

Water management in different civilisations

Water management has been a crucial element in the development of ancient civilisations, determining their prosperity and survival throughout history. Different cultures developed unique and sophisticated systems for water development, storage and distribution, adapted to their specific geographic conditions and needs (Mithen, 2012).

In Mesopotamia, considered the cradle of civilisation, water management focused on harnessing the Tigris and Euphrates rivers. The Sumerians and Babylonians developed complex systems of canals and dykes to control seasonal floods and direct water to agricultural fields. The Code of Hammurabi, one of the earliest known legal codes, included specific regulations on water management, highlighting its social and economic importance (Mays, 2010).

Ancient Egypt based its prosperity on the management of the Nile and its annual floods. The Egyptians developed a sophisticated system of nilometers to measure and predict the river's floods, along with a network of canals and dykes to harness water and fertile sediments. This system not only facilitated agriculture, but also determined taxation and the social organisation of the empire (Hassan, 2014).

The Roman civilisation took hydraulic engineering to new heights with the construction of monumental aqueducts, thermal baths and urban distribution systems. The Romans implemented advanced technologies such as lead pipes, control valves and filtration systems. Their administrative system included specific officials (curator aquarum) in charge of overseeing the hydraulic infrastructure and regulating its use (Hodge, 2013).

In the Americas, the Maya developed innovative systems for water collection and storage in regions lacking permanent rivers. They built artificial reservoirs (aguadas), rainwater harvesting systems and underground canals. The Aztecs created chinampas, highly productive artificial islands in shallow lakes, and developed dike systems to control water levels in the city of Tenochtitlan (Lucero & Fash, 2019).

Chinese civilisation implemented sophisticated flood control techniques and irrigation systems, including the Grand Canal, the most extensive waterworks in the ancient world. The development of the Dujiangyan irrigation system in the 3rd century BC transformed the Min River basin into one of the most productive agricultural regions in China, and it continues to operate to this day (Needham & Wang, 2015).

In ancient India, water management incorporated religious and cultural elements along with advanced technical solutions. Systems of step wells (baolis), storage tanks and distribution canals were integrated into urban and religious planning. The Arthashastra treatise included detailed guidelines on water management and its importance for governance (Shaw, 2018).

Arab civilisations developed innovative systems for managing water in arid environments, including qanats (underground canals) and sophisticated urban distribution systems. Medieval Arab engineers perfected mechanical devices for lifting water and wrote important treatises on hydraulics (Al-Hassan & Hill, 2016).

The Incas in South America built complex systems of agricultural terraces and irrigation canals in mountainous terrain, demonstrating a deep understanding of hydraulics and engineering. Their water distribution systems included ceremonial fountains and drainage systems that are still admired today for their technical precision (Ortloff, 2017).

Hydraulic solutions that last

Many of the hydraulic solutions developed by ancient civilisations continue to function or influence modern water management systems, demonstrating the sophistication and durability of these ancient designs. These structures have not only stood the test of time, but also offer valuable lessons for addressing today's water management challenges (Smith, 2019).

The Dujiangyan irrigation system in China, built in 256 BC, is still operational and provides water to more than 50 million people in Sichuan province. Its design, which combines flood control with irrigation water distribution, uses hydraulic principles that are still relevant in modern engineering. The system does not require permanent dams and takes advantage of the natural topography to divide and regulate water flow, minimising the maintenance required (Wang, 2020).

Middle Eastern qanats, underground tunnels that transport water by gravity from aquifers to the surface, are still in operation in countries such as Iran, Oman and Afghanistan. These systems, which date back more than 2000 years, provide drinking and irrigation water without the need for mechanical pumping, representing a sustainable solution for arid regions. They are designed to minimise evaporation and maintain water quality, which are crucial in the context of current climate change (Wilson, 2018).

Inca agricultural terraces in the Peruvian Andes continue to be used by local communities. Their design incorporates drainage and erosion control systems that have prevented landslides for centuries. The construction principles used in these terraces are currently being studied to develop sustainable solutions for mountain agriculture and soil conservation (Ortloff, 2021).

Traditional water tanks in India, known as kunds or baolis, are still relevant for rural water management. These rainwater harvesting and storage systems have inspired modern water harvesting and aquifer recharge projects. Their design allows for natural filtration and maintenance of water quality, aspects that are incorporated in contemporary water management systems (Kumar, 2017).

Roman aqueducts, although largely disused for their original purpose, have significantly influenced the design of modern water distribution systems. The hydraulic principles used in their construction, such as the use of gravity and pressure control, remain fundamental to hydraulic engineering today. Some aqueducts, such as the Segovia aqueduct in Spain, maintained their function until the 20th century (Mays, 2020).

The Aztec chinampas have inspired modern systems of urban agriculture and sustainable water management. This farming method, which combines water management with intensive agricultural production, offers solutions for sustainable urban agriculture and wetland management. Contemporary projects in Mexico and other countries are adapting these principles to develop resilient agricultural systems (Martinez, 2016).

Flood control systems developed in the Netherlands over centuries continue to evolve and adapt to new challenges. The basic principles of water management established during the medieval period are still relevant, although they have been modernised with today's technology. The concept of 'living with water' rather than fighting against it has influenced modern climate change adaptation strategies (Van der Veer, 2019).

The importance of these historic water solutions lies not only in their continued functionality, but also in the sustainable principles they embody. In an era of increasing water stress and climate change, these traditional technologies offer valuable lessons in adaptability, sustainability and resilience in water management.

CAPITULO IV

LECCIONES DEL PASADO VISIÓN DEL FUTURO

Durability analysis of old structures

Ancient structures, such as the Great Pyramid of Giza and the Roman Colosseum, have endured over the centuries, defying not only the elements but also human intervention. This phenomenon is not only a testament to the skill of their builders, but also to the careful choice of materials and construction techniques.

The Great Pyramid, built more than 4,500 years ago, is a paradigmatic example. Its geometric design and the quality of the materials used - limestone and granite - have allowed this monumental work to survive for millennia. The choice of these materials was not accidental; their resistance to erosion and their ability to withstand massive loads are characteristics that the ancient Egyptians understood and exploited. The pyramid is precisely aligned with the cardinal points, which also reflects a deep astronomical and geographical understanding.

A detailed analysis of these structures reveals fundamental principles of durability: the use of stable geometries, the choice of local materials and an understanding of the environment. For example, the Roman Colosseum was not only designed to be a monumental space for public performances, but its elliptical structure and use of the arch allowed for the efficient distribution of loads, which has contributed to its longevity. The combination of concrete, travertine stone and brick in its construction provides exceptional strength. These lessons are crucial for modern engineers seeking to create buildings that are not only functional, but also stand the test of time.

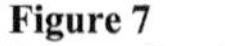

Figure 7

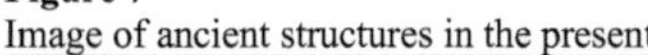

Image of ancient structures in the present

Note. Resource adapted by ideogram

Sustainable building techniques of the past

Ancient civilisations implemented sustainable techniques that are relevant today. The Romans designed their buildings with rainwater harvesting systems and natural ventilation techniques that minimised the need for artificial heating and cooling. This approach was not only energy efficient, but also reflected a deep respect for the environment.

Vernacular architecture in various cultures shows an ingenious use of local resources. In arid regions, houses were built with thick walls to keep temperatures cool in summer and warm in winter. This type of design can be seen in the traditional houses of the American Southwest, where adobe was used to regulate the interior temperature. Reuse of materials and adaptive design are practices that can be rediscovered and applied in modern construction to reduce the ecological footprint.

A notable example is the use of adobe in many indigenous cultures. This natural material is not only abundant and easy to work with, but also has excellent thermal properties. Buildings made of adobe naturally regulate the interior temperature, reducing dependence on mechanical systems for heating or cooling.

Figure 8
Image of sustainable construction linked to the present day

Note. Resource adapted by ideogram

Restoration and preservation methods

The restoration and preservation of historic monuments require a meticulous approach that respects both aesthetics and structural integrity. The restoration of the Roman Colosseum is an emblematic case where modern techniques have been used to stabilise the structure without compromising its authenticity. This includes the use of compatible materials that mimic the original properties without altering the historic character of the monument.

In addition, traditional methods are fundamental in this process. In many cultures, ancient techniques have been used to repair stone walls, ensuring that the interventions are invisible to the naked eye and maintaining the cultural value of the building. For example, in Europe, many Gothic cathedrals have been restored using medieval techniques to ensure that repairs are consistent with the original structures.

Preservation is not only about restoration; it is also about adapting these spaces to new functions without losing their historical essence. A contemporary example is the adaptive use of old factories in cultural or residential spaces, where the original structure is preserved while incorporating modern elements.

Figure 9

Image preservation and restoration

Note. Resource adapted by ideogram

Application of ancient principles in modern engineering

The architectural principles used by ancient engineers can offer innovative solutions to contemporary challenges. The use of the arch and vault in Roman buildings not only provided exceptional structural strength, but also allowed for the creation of large, light-filled spaces. These concepts can be adapted to design modern buildings that are more efficient in terms of space and resources.

In addition, bioclimatic design principles used by ancient civilisations - such as proper orientation to maximise sunlight or minimise wind - are now more relevant than ever in a context where sustainability is a global priority. Integrating these principles into contemporary design can result in buildings that are more resilient and less dependent on external energy sources.

For example, traditional Japanese houses use pitched roofs to manage rainfall and prevent unwanted accumulation of snow or water. This type of design can be critical today when considering the changing climatic conditions caused by global warming.

Contemporary challenges and historical solutions

Current challenges in civil engineering include climate change, urban sprawl and natural resource scarcity. Historical solutions offer a rich source of inspiration for addressing these problems. For example, Roman aqueducts were not only ingenious in their design; they were also early examples of efficient water management, which is crucial in a world where this resource is increasingly scarce.

The techniques used by civilisations such as the Incas to manage their hydraulic systems can be adapted to meet today's water-related challenges. Implementing sustainable systems based on these historical principles can contribute significantly to more effective environmental management.

A contemporary case is the traditional Andean system known as "andenes", where agricultural terraces are created to allow better water control and higher agricultural productivity in mountainous terrain. Such practices can be key to addressing challenges such as desertification or prolonged droughts.

The future of resilient construction

The future of the building sector will depend to a large extent on our ability to merge the lessons of the past with contemporary technological innovations. Resilience to natural disasters must be a fundamental priority in modern architectural design. This means not only using durable materials and time-tested techniques, but also developing infrastructure that can adapt to changing climatic conditions.

Incorporating modular and flexible design can enable buildings to better respond to extreme events such as earthquakes or floods. In addition, promoting a building culture that values both innovation and respect for traditions can result in more effective and sustainable solutions.

One promising approach is the growing use of digital technologies such as 3D modelling and 3D printing to create customised structures that respond to specific needs while optimising resources. These technologies can facilitate more efficient and inclusive design by enabling greater community participation in construction processes.

Lessons for the new generation of engineers

It is essential that new generations of engineers understand and appreciate the legacy left by their predecessors. Civil engineering education must integrate studies of ancient techniques alongside modern technologies. Fostering a critical mindset towards learning from the past will enable future engineers to develop innovative solutions that are both effective and sustainable.

Incorporating interdisciplinary studies - including history, ecology and sociology - will enrich their education and enable them to approach complex problems from multiple angles. In doing so, they will be better prepared to face the challenges of the future while honouring the lessons learned over time.

It is equally important to foster a professional ethic focused on social and environmental responsibility. Engineers must be aware of the impact their decisions have on local communities and ecosystems. This involves not only designing safe and functional buildings, but also considering how they interact with their social and environmental context.

REFERENCIAS

Adam, J. P. (2016). Roman building: Materials and techniques (4th ed.). Routledge.

Agencia Andina (2018). The wisdom that emanates from Machu Picchu: a study reveals the seismic-resistant engineering of the Incas.https://portal.andina.pe/edpespeciales/2018/machupicchu/index.html

Ahunbay, Z. (2016). Conservation of Byzantine Istanbul. *Journal of Architectural Conservation*, 22(1), 4-20.

Archaeological and Anthropological Sciences,*12*(4),1-15. Management History,* 28*(3),156-178

Alder, K. (2019). *Engineering the revolution: Arms and enlightenment in France, 1763-1815.* Princeton University Press.

Altman, N. (2018). *Sacred water: The spiritual source of life*. Inner Traditions.

Bardill, J. (2017). *The monuments of ancient Constantinople*. Cambridge University Press.

Bassett, S. (2015). The urban image of late antique Constantinople. Cambridge University Press.

Beckmann, J. (2021). *A history of inventions, discoveries, and origins* (W. Johnston, Trans.). Cambridge University Press.

Browne, M. W. (2017). *Ancient construction techniques: From pyramids to cathedrals*. Oxford University Press.

Çeçen, K. (2014). The longest Roman water supply line. Turkish Academy of Sciences.

Cemex Ventures(n.a.) Sustainable building materials.https://www.cemexventures.com/es/materiales-construccion-sostenib

Buildable(n.a.) Sustainable construction.https://www.construible.es/construccion-sostenible

Cerlalc(n.d.) Looking at the past, thinking and intervening in the present,... planning for the future: Part 1.https://cerlalc.org/mirar-el-pasado-pensar-e-intervenir-el-presente-planificar-el-futuro-parte-1/

Crow, J., Bardill, J., & Bayliss, R. (2008). The water supply of Byzantine Constantinople. Society for the Promotion of Roman Studies.

Crow, J. (2012). Water and late antique Constantinople. In L. Grig & G. Kelly (Eds.), *Two Romes: Rome and Constantinople in late antiquity* (pp. 116-135). Oxford University Press.

Cool Culture.(2024,October 22) Roman Colosseum: history, features and fun facts.https://www.culturagenial.com/es/coliseo-romano/

Dark, K., & Harris, A. (2008). The last Roman: Britain and the Roman inheritance. Tempus.

Davidson, R. (2021). Stone age revolution: The transformation of building techniques. *Construction History Review*, 18(2), 45-62.

DeLaine, J. (2021). The baths of Caracalla: A study in the design, construction, and economics of large-scale building projects in imperial Rome.* Journal of Roman Archaeology*.

Downey, G. (2020). *Ancient Egyptian construction and architecture* (2nd ed.). Dover Publications.

Sustainable Economy.(2024,October 22) The secret of the Great Wall of China's resistance to erosion has been revealed. https://economiasustentable.com/noticias/revelaron-el-secreto-de-la-resistencia-contra-la-erosion-de-la-gran-muralla-china/

El Tiempo.(2024,October 22) How were the pyramids of Egypt made? This is what artificial intelligence says.https://www.eltiempo.com/cultura/gente/como-se-hicieron-las-piramides-de-egipto-esto-dice-la-inteligencia-artificial-3391603

Evans, R. M. (2018). *The ancient engineers: Technology and invention from the earliest times to the Renaissance*. MIT Press.

Fagan, B. M. (2021). *The first civilizations: Ancient Mesopotamia and early Egypt* (3rd ed.). Thames & Hudson.

Faster Capital(n.d.) The Great Wall of China as a defence system.https://fastercapital.com/es/tema/la-gran-muralla-china-como-sistema-de-defensa.htm

Fagan, B. M. (2021). *The first civilizations: Ancient Mesopotamia and early Egypt* (3rd ed.). Thames & Hudson.

García,J.(2016). Chapter 4: Lessons from the past for the future. in [Title of the book or document if available]. Academia.edu.https://www.academia.edu /31970516 /Chapter_4_Lessons_from_the_Past_for_the_Future.

García-Martínez, A. (2022). From wood to stone: Architectural transformations in antiquity.* Revista de Historia de la Construcción,* 15(4), 78-95.

Guerrero Ortiz,L.(2019,November11) Vision of the future and understanding the past.Instituto Educacción.https://institutoeducaccion.org/vision-de-futuro-y-comprension-del-pasado/

Hodge, A.T.(2002) * Roman aqueducts & water supply * Bristol Classical Press.

Hodge, A.T.(2017).* Roman aqueducts & water supply* (2nd ed.). Bristol Classical Press.

INUE.(2024,October 22) Historical Architecture: Roman Coliseum.https://inue.edu.mx/blog/arquitectura-histrica-coliseo-romano

Infobae.(2023,September 1) Earthquake engineering of the Incas: connection between architecture and geological risk in Peru. https://www.infobae.com/peru/2023/09/01/la-ingenieria-sismica-de-los-incas-conexion-entre-la-arquitectura-y-el-riesgo-geologico-en-el-peru/

Johnson,M., & Lee,S.(2020).* Early construction materials and their applications.* Archaeological Studies Quarterly,* 28(1), 23-41.

Kostof,S.(2023).* A history of architecture: Settings and rituals* (4th ed.). Oxford University Press.

Lancaster,L.C.(2021).* Concrete vaulted construction in Imperial Rome: Innovations in context.* Cambridge University Press.

Lendering,J.(2021).* Cities of ancient Mesopotamia.* Routledge.

Lucero,L.J., & Fash,B.W.(2019).* Water and ritual: The rise and fall of classic Maya rulers.* University of Texas Press.

Malissard,A.(2001).* The Romans and Water: The Culture of Water in Ancient Rome.* Herder.

Matthewson,C.(2019).* Historic bridges: Evaluation, preservation, and management.* McGraw-Hill Education.

Mays,L.W.(2010).* Ancient water technologies.* Springer.

Mithen,S.(2012).* Thirst: Water and power in the ancient world.* Harvard University Press.

Moropoulou,A., Karoglou,M., & Delegou,E.T.(2020).* Ancient building technologies and materials.* Archaeological and Anthropological Sciences,*12*(4),1-15.

National Geographic.(2022,May 24) The Pyramids of Giza: Clues to Construction and Myths. https://www.nationalgeographic.es/historia/piramides-guiza-claves-construccion-mitos

National Geographic.(2024,October 22) Study reveals secret of Roman concrete's strength. https://historia.nationalgeographic.com.es/a/un-estudio-revela-el-secreto-de-la-resistencia-del-hormigon-romano_18951

National Geographic.(2024,October 22) The Great Wall of China: the world's greatest engineering feat. https://historia.nationalgeographic.com.es/a/gran-muralla-china-mayor-obra-ingenieria-mundo_8272

Needham,J., & Wang,L.(2015).* Science and civilisations in China: Volume 4 ,Physics and physical technology ,Part 3,Civil engineering and nautics.* Cambridge University Press}

Nuttgens,P.(2021).* The story of architecture: From antiquity to the present* (3rd ed.). Phaidon Press

Pagès,J., & Santisteban,A.(2009). Learning the past, understanding the present and building the future: lessons from a research on the history taught.XII Jornadas Interescuelas / Departamentos de Historia.Facultad de Humanidades y Centro Regional Universitario

Bariloche.Universidad Nacional del Comahue. https://cdsa.aacademica.org/000-008/843.pdf

Park,C.(2014).* Traditional irrigation techniques in different cultures.* Moropoulou,A., Karoglou,M., & Delegou,E.T.(2020).* Ancient building technologies and materials.

Proyecpro(n.d.) Sustainable construction in Latin America.https://proyecpro.com/construccion-sostenible-en-latinoamerica/

Quilosa(2020,september17) Sustainable construction techniques.https://quilosa.com/eficiencia-energetica/tecnicas-construccion-sostenible/

Ritti,T., Grewe,K., & Kessener,P.(2017) A relief of a water-powered stone saw mill on a sarcophagus at Hierapolis and its implications.* Journal of Roman Archaeology,*20(1),138-163.

Roberts,C., & Chen,L.(2023).* Social implications of construction material evolutio

Scarborough,V.L .(2016) The flow of power: Ancient water systems and landscapes.SAR Press.

Shaw,B.D .(2020) Bringing in the sheaves: Economy and metaphor in ancient Rome.University of Toronto Press.

Smith,J.R .(2019) Enduring solutions: Historical water systems in modern context.Water Engineering and Management,*28*(4),112-126.

Smith,M.E .(2019) Cities: The first 6,000 years.Viking Press

Trevor,H.(2016).* Engineering in the ancient world.* University of California Press.

Taylor,R .(2020) Roman builders: A study in architectural process.Cambridge University Press.

Tickets Rome(n.d.) History of the Colosseum in Rome (Flavian amphitheatre).https://www.tickets-rome.com

Thompson,J .(2018) Greek and Roman civil engineering.Thames & Hudson

Vitruvius,P .(2019) The ten books on architecture(M.H.Morgan ,Trans.).Dover Publications

We Collect Postcards.(2024,October 22) Fun facts about the Pyramids of Giza: 15 surprising facts.https://wecollectpostcards.com/curiosidades-piramides-de-giza/

Westbrook,N.(2019). The architecture of the great palace of Constantinople. In P.Magdalino & N.Ergin(Eds.), Constantinople: Archaeology of a Byzantine megapolis(pp. 201-242). Oxbow Books.

Wilson,A .(2008) Hydraulic engineering and water supply.In J.P.Oleson(Ed.), Oxford handbook of engineering and technology in the classical world(pp285-318).Oxford University Press

Wilson,A .(2018) Hydraulic engineering and water supply.In The Oxford handbook of engineering and technology in the classical world(pp285-318).Oxford University Press

Wilson,A .(2019) Engineering and technology in the classical world.Oxford University Press.

Wilkinson,T.J .(2013) Archaeological landscapes of the Near East.University of Arizona Press.

Wikipedia(n.d.) Navigation channel.https://es.wikipedia.org/wiki/Canal_de_navegaci%C3%B3n

Wikipedia(n.d.) Puerto.https://es.wikipedia.org/wiki/Puerto

Zhang,Y., Wu,X., & Li,H.(2023) Limitations and advantages of traditional building materials.International Journal of Construction History,*40*(2),123-140

Printed by Books on Demand GmbH, Norderstedt / Germany